AMAZING

ARCHITECTURE
2025 CALENDAR

JANUARY 2025

SUNDAY	MONDAY	TUESDAY	WEDNESDAY	THURSDAY	FRIDAY	SATURDAY
29	30	31	1	2	3	4
5	6	7	8	9	10	11
12	13	14	15	16	17	18
19	20	21	22	23	24	25
26	27	28	29	30	31	1

FEBRUARY 2025

SUNDAY	MONDAY	TUESDAY	WEDNESDAY	THURSDAY	FRIDAY	SATURDAY
26	27	28	29	30	31	1
2	3	4	5	6	7	8
9	10	11	12	13	14	15
16	17	18	19	20	21	22
23	24	25	26	27	28	1

MARCH 2025

SUNDAY	MONDAY	TUESDAY	WEDNESDAY	THURSDAY	FRIDAY	SATURDAY
23	24	25	26	27	28	1
2	3	4	5	6	7	8
9	10	11	12	13	14	15
16	17	18	19	20	21	22
23 / 30	24 / 31	25	26	27	28	29

APRIL 2025

SUNDAY	MONDAY	TUESDAY	WEDNESDAY	THURSDAY	FRIDAY	SATURDAY
30	31	1	2	3	4	5
6	7	8	9	10	11	12
13	14	15	16	17	18	19
20	21	22	23	24	25	26
27	28	29	30	1	2	3

MAY 2025

SUNDAY	MONDAY	TUESDAY	WEDNESDAY	THURSDAY	FRIDAY	SATURDAY
27	28	29	30	1	2	3
4	5	6	7	8	9	10
11	12	13	14	15	16	17
18	19	20	21	22	23	24
25	26	27	28	29	30	31

JUNE 2025

SUNDAY	MONDAY	TUESDAY	WEDNESDAY	THURSDAY	FRIDAY	SATURDAY
1	2	3	4	5	6	7
8	9	10	11	12	13	14
15	16	17	18	19	20	21
22	23	24	25	26	27	28
29	30	1	2	3	4	5

JULY 2025

SUNDAY	MONDAY	TUESDAY	WEDNESDAY	THURSDAY	FRIDAY	SATURDAY
29	30	1	2	3	4	5
6	7	8	9	10	11	12
13	14	15	16	17	18	19
20	21	22	23	24	25	26
27	28	29	30	31	1	2

AUGUST 2025

SUNDAY	MONDAY	TUESDAY	WEDNESDAY	THURSDAY	FRIDAY	SATURDAY
27	28	29	30	31	1	2
3	4	5	6	7	8	9
10	11	12	13	14	15	16
17	18	19	20	21	22	23
24 / 31	25	26	27	28	29	30

SEPTEMBER 2025

SUNDAY	MONDAY	TUESDAY	WEDNESDAY	THURSDAY	FRIDAY	SATURDAY
31	1	2	3	4	5	6
7	8	9	10	11	12	13
14	15	16	17	18	19	20
21	22	23	24	25	26	27
28	29	30	1	2	3	4

OCTOBER 2025

SUNDAY	MONDAY	TUESDAY	WEDNESDAY	THURSDAY	FRIDAY	SATURDAY
28	29	30	1	2	3	4
5	6	7	8	9	10	11
12	13	14	15	16	17	18
19	20	21	22	23	24	25
26	27	28	29	30	31	1

NOVEMBER 2025

SUNDAY	MONDAY	TUESDAY	WEDNESDAY	THURSDAY	FRIDAY	SATURDAY
26	27	28	29	30	31	1
2	3	4	5	6	7	8
9	10	11	12	13	14	15
16	17	18	19	20	21	22
23 / 30	24	25	26	27	28	29

DECEMBER 2025

SUNDAY	MONDAY	TUESDAY	WEDNESDAY	THURSDAY	FRIDAY	SATURDAY
30	1	2	3	4	5	6
7	8	9	10	11	12	13
14	15	16	17	18	19	20
21	22	23	24	25	26	27
28	29	30	31	1	2	3

SOME OF THE MOST FAMOUS ARCHITECTS IN THE LAST 100 YEARS

Frank Lloyd Wright
Le Corbusier
Mies van der Rohe
Zaha Hadid
Frank Gehry
Antoni Gaudí
I.M. Pei
Norman Foster
Renzo Piano
Santiago Calatrava
Louis Kahn
Eero Saarinen
Bjarke Ingels
Tadao Ando
Richard Meier
Peter Zumthor
Oscar Niemeyer
Alvar Aalto
Philip Johnson
Jean Nouvel
Rem Koolhaas
Moshe Safdie
Daniel Libeskind
Kengo Kuma
Shigeru Ban
David Chipperfield
Kenzo Tange
Steven Holl
Robert Venturi
Michael Graves

SOME OF THE BEST BUILDINGS IN THE LAST 100 YEARS

Fallingwater (1935) - Frank Lloyd Wright
Villa Savoye (1931) - Le Corbusier
Guggenheim Museum (1959) - Frank Lloyd Wright
Sydney Opera House (1973) - Jørn Utzon
Seagram Building (1958) - Mies van der Rohe
TWA Flight Center (1962) - Eero Saarinen
Pompidou Centre (1977) - Renzo Piano and Richard Rogers
Louvre Pyramid (1989) - I.M. Pei
Sagrada Família (ongoing, begun 1882) - Antoni Gaudí
Guggenheim Museum Bilbao (1997) - Frank Gehry
Burj Khalifa (2010) - Adrian Smith (Skidmore, Owings & Merrill)
Heydar Aliyev Center (2012) - Zaha Hadid
Taj Mahal Palace Hotel (2004 restoration) - W.A. Chambers, Charles Stevens
30 St Mary Axe (The Gherkin) (2004) - Norman Foster
National Congress of Brazil (1960) - Oscar Niemeyer
Marina Bay Sands (2010) - Moshe Safdie
Petronas Towers (1998) - César Pelli
Walt Disney Concert Hall (2003) - Frank Gehry
One World Trade Center (2014) - David Childs (Skidmore, Owings & Merrill)
The Shard (2012) - Renzo Piano
MAXXI Museum (2009) - Zaha Hadid
Institute of Contemporary Art (2006) - Diller Scofidio + Renfro
Millennium Bridge (2000) - Norman Foster
Beijing National Stadium (Bird's Nest) (2008) - Herzog & de Meuron
Lloyd's Building (1986) - Richard Rogers
JFK Presidential Library (1979) - I.M. Pei
Kimbell Art Museum (1972) - Louis Kahn
Museum of Islamic Art (2008) - I.M. Pei
Casa da Música (2005) - Rem Koolhaas
Seattle Central Library (2004) - Rem Koolhaas and Joshua Prince-Ramus